"I do seriously, and upon good grounds, affirm it possible to make a flying chariot. In which a man may sit, and give such a motion unto it, as shall convey him through the air. And this perhaps might be made large enough to carry many men at the same time together with food for their trip.

"The perfecting of such an invention, would be of such excellent use, that it would be enough, not only to make a man famous, but the age also wherein he lives.

"So that notwithstanding all these seeming impossibilities, it is likely enough that there may be a means invented of journeying to the Moon; and how happy shall they be, that are first successful in this attempt?"

Bishop John Wilkins
1640

We acknowledge the financial support of the Government of Canada through the Book
Publishing Industry Development Program for our publishing activities.

Published by Apogee Books, Box 62034, Burlington,
Ontario, Canada, L7R 4K2, http://www.apogeebooks.com
Tel: 905 637 5737

Printed and bound in Canada

Apollo 11 - First Men on the Moon by Robert Godwin
ISBN 1-894959-27-2

©2005 Robert Godwin

First Men on the Moon:

APOLLO 11

by Robert Godwin

THE LONG ROAD TO TRANQUILITY BASE

It is entirely possible that when human civilization has reached its zenith and finally slipped into obscurity the footprints at Tranquility Base will remain unchanged. The only legacy of a stubborn species that once strived for greatness.

Our quest to understand the nature of ourself and our place in the universe is as old as human memory. Before the ancient Greek philosophers wrote of these profound matters there were the Babylonians and the astronomers of the Indian sub-continent who speculated without the benefit of technology. They believed that the answer to understanding who we are lay in the abstract depths of heaven. These early observers passed on their ideas, down through the generations, until they were ultimately absorbed by their cousin-descendants in Greece in the first millenium BCE.

In the last hundred years we have begun to realize that we hold much more in common with the universe at large than our ancestors ever realized. The metaphysically inclined amongst us have always believed that the path to enlightenment lies in the heavens, while those of us who are more inclined to the physical have learned that indeed we are more closely connected with the heavens than we could have ever foreseen. Whether you believe in a divine creation from above or you believe that we are an accidental collection of heavy elements created in the heart of a long dead exploding star you are left with the feeling that the answer is "out there."

For most of recorded history mankind believed that the universe revolved around the Earth. It was taken for granted in almost all of the world's early civilizations that the sun, moon and stars revolved overhead, and the Earth stood still.

The erratic movements of the planets were explained away by attributing them to the wanderings of the gods. Many early societies worshipped the sun and the moon and so, inevitably, with the birth of primitive astronomy, the planets were promoted to divine status.

The spokes of ancient Greek astronomy revolved firmly around a philosopher named Thales of Miletus who after much careful observation was able to accurately predict a solar eclipse on May 28th 585 BCE. It was the first documented scientific experiment in astronomy. For the next two hundred years this highly charged debate continued until the curator of the great library in Alexandria suggested that the Earth rotated around the Sun. His name was Aristarchus of Samos and he was swiftly condemned for making such a radical suggestion.

However, Aristarchus' idea would linger quietly in ancient scrolls and would ultimately be translated into Latin almost a thousand years later. These translations would find their way into central Europe and eventually onto the desk of a Polish priest called Nicolaus Copernicus.

Copernicus would spend over thirty years of his life recording the movements of the heavens and came to the conclusion that the Earth was not the centre of the universe and that the planets, including the Earth, did in fact rotate around the Sun. The only exception was the Moon, which he believed was a world unto itself that rotated around the Earth. He was so concerned that he might face the same persecution as his ancient predecessor, Aristarchus, that he didn't publish his findings. His theories would not be revealed until after his death in 1543.

Sixty six years later another priest, Johanes Kepler, proved that Copernicus was correct. Kepler refined the mathematics and showed that the discrepancies apparent in Copernicus' observations were explainable if the planets and the moon moved, not in circular, but elliptical orbits.

The following year three different men simultaneously submitted applications to the Dutch government for the right to exclusively manufacture an instrument for seeing things "at a distance". It would become known as a *telescope* and in January of 1610 a brilliant philosopher/scientist named Galileo Galilei, hearing of this magnificent invention, built his own and turned it toward the heavens.

Galileo's celebrated moment would create the single greatest shift in the comprehension of the universe in the history of humankind. When he pointed his primitive telescope towards the Milky Way he could see that it was a band of incalculable numbers of stars. When he looked at the moon he could see mountains, valleys and blemishes of the kind that covered the Earth, irrefutably the moon was a world unto itself. Most importantly he turned it toward Jupiter and what he saw were four small moons circling the giant planet. Undeniably, the Earth was not the centre of the universe and anyone with a telescope could see this with their own eyes. Galileo proudly announced this fact in his book *The Sidereus Nuncius* and was swiftly reprimanded by those who continued to cling to the Earth-centred universe sanctioned centuries earlier by Aristotle.

Even though there were many who would silence Kepler and Galileo, the truth is hard to shroud once it has been revealed. A larger universe had been exposed and it was no longer possible to conceal it behind rhetoric and superstition.

Almost three decades later two English bishops, with widely differing motives, each published a book. The first was a science fiction novel called *The Man in the Moone*, and it told a story of a Spanish adventurer who flies to the moon aboard a mechanical device pulled aloft by a flock of geese. The author, Bishop Francis Godwin, was too concerned with his reputation to publish the book so it would not be released until five years after his death. The subtext of the story would directly inspire a work of science by Bishop John Wilkins who the same year, 1638, would publish his book *Discovery of a New World, or, a Discourse Tending to prove, that 'tis Probable there may be another Habitable World in the Moon*.

Wilkins would present thirteen propositions to suggest that the moon was a real world, possibly populated, and that it may be feasible to travel there and do commerce with the inhabitants. It was a radical proposal which would surely have found Wilkins in a jail cell had he published it outside of England. A quote from Wilkins can be seen on the title page of this book.

Wilkins was a member of the distinguished Royal Society and was one of the great intellectual stars of his era. Forty eight years later another member of that illustrious scientific society would once more open our eyes to a larger universe, it was 1686 and his name was Sir Isaac Newton. His book was called *Philosophiae Naturalis Principia Mathematica* and it would cause a shift in humankind's understanding of the universe on a similar scale to that caused by Galileo.

Working from decades of careful observations, made by the likes of Kepler and Galileo, Newton showed how the movements of the universe could be defined mathematically. His extraordinary book meticulously explained the orbits of the moon and planets. The concept of gravity was first articulated in detail in the *Principia*. Newton's theory finally explained how people could be standing on any part of the Earth's surface while their feet were always planted firmly facing its center.

The movements of the moons of Jupiter could also be conveniently explained by Newton's mathematical predictions and so his finding was dubbed the *Universal* law of Gravitation. Joining the ranks of Galileo, Copernicus and Kepler, Newton would be forever immortalized by his remarkable achievement. His mathematical model of the universe would stand intact for the next two centuries and would finally bring closure to the arguments about the order of the solar system. The moon and planets were definitely real destinations that behaved in predictable ways. Newton's principles would also apply to any object attempting to leave planet Earth.

Almost a hundred years would pass, during which Newton's new philosophy would stir contentious debate, then another blow would be struck against superstition. Since the earliest recordings of mankind the solar system had been limited to just six planets. Then the science of studying the heavens reached an unexpected new peak thanks to a naturalized German resident of the city of Bath, in England. On the night of March 13th 1781 William Herschel would become the founding member of the most exclusive club in science. He would discover a new planet.

Herschel's discovery would mark the end of the era of the immutable universe. Although most scientists had accepted the notion of a much larger (perhaps infinite) universe the discovery of a new planet, which was barely visible to the untrained eye, was a watershed event in the history of cosmology. It muted the voices of those who argued that somehow everything in the sky was there for man's direct benefit. Logically, if you couldn't even see this new planet then how could it possibly have any bearing on mankind's destiny? Astrology would never be the same.

Two years after Herschel's discovery two French paper-makers stepped into the limelight and earned a place in the history books. On June 5th 1783 the Montgolfier brothers, Etienne and Joseph, invited an audience to the marketplace of Annonay to witness the ascent of the world's first flight of a hot air balloon. Three months later, under royal decree, they repeated the flight but this time carrying aloft an assortment of barnyard animals. Then a month later the first human to leave the Earth by means of a controlled artificial device took to the air. His name was Francois Pilâtre de Roziers and he flew in the Montgolfier balloon for four minutes and 25 seconds at an altitude of 84 feet before returning safely to the Earth.

For the next 120 years flight was the sole domain of the balloonists but it could hardly be said that they ruled the skies. Balloons were hard to steer and were subject to the vagaries of weather and wind.

In 1865 a French legal student, ensconced in the Parisian arts community, took it upon himself to propose another method of high speed travel. His name was Jules Verne and his books *From the Earth to the Moon* and *Around the Moon* suggested the possibilities of ballistic flight. Verne would place his aviators inside a cannon shell and blast them from a giant gun on a trip to the moon. Although Verne's story was meant to be satirical it sparked a discussion in many corners of the world about the feasibility of such a voyage. Science was catching up with fantasy.

One of Verne's avid readers was a half-deaf school teacher living in Kaluga, Russia. His name was Konstantin Tsiolkovsky and he had spent his life reading and learning about science. On March 28th 1883 he wrote, "*Consider a cask filled with a highly compressed gas. If we open one of its taps the gas will escape through it in a continuous flow, the elasticity of the gas pushing its particles into space will also continuously push the cask itself. The result will be a continuous motion of the cask.*" He also postulated that if the cask was to have multiple taps they could be regulated in such a way to make it possible to actually *steer* the cask. This was not the invention of the rocket—that moment was lost somewhere back in 12th century China—rather this was the realization that a rocket would work in space. This was a crucial discovery since over fifty years later many scientists would continue to believe that a rocket needed air to be able to create thrust. These ill-informed sceptics were floundering under a fundamental misunderstanding of Newton's principles—the idea that a rocket's exhaust needed something to push against. Tsiolkovsky knew better and he set about proving it mathematically.

By the end of the 19th century the flow of information was accelerated by the opening up of the American wilderness. Wood pulp was becoming readily accessible and this new plenitude caused a surge in the available reading materials. The general public were inundated with magazines and newspapers. The perfection of the process for refining aluminum also accelerated the mass reproduction of photographs. Into this new flood of printed information came a torrent of news about the heavens.

Astronomers in Italy, England, France and America, using ever more powerful telescopes, discovered moons around Mars and thought they saw enormous parallel lines on the Martian surface. The debate swirled around the globe, enrapturing all walks of society. Did Mars have an ancient civilization? Then in 1897 a renowned English writer of fiction, called Herbert George Wells, unleashed a story of how an utterly alien life form might invade the Earth. In Germany a less sceptical author by the name of Kurd Lasswitz proposed invasion by a more advanced and benevolent form of Martian who used space stations as a staging post for their Earthly intrusions. These three authors, Wells, Lasswitz and Verne would inspire a whole generation of young scientifically inclined enthusiasts.

Besides Tsiolkovsky, a young ex-soldier from Transylvania called Hermann Oberth and a frail mild-mannered scholar from New England called Robert Goddard, would both devour these fantastic fictions of space travel. But before any progess could be made in the real world it was clear that someone would need to improve on the balloon.

Many adventurers believed that it was possible to build a mechanical device that could overcome the inherent inadequacies of the balloon but it would not be until December of 1903 that two bicycle mechanics from Ohio would truly devise a method for conquering the skies.

The Wright Flyer was a marvel of home engineering. The key innovation was how the wings could be warped through an intricate mechanism that connected to the rudder. At 10 a.m. on the morning of December 17th 1903 a 21 mph wind was blowing off-shore at Kitty Hawk on the North Carolina coast. Orville Wright climbed aboard the delicate structure of wood and wire and waited for his moment with destiny. At 10.30 he coaxed the fragile machine into the air for a few seconds and it traveled just over 120 feet. It was not the first time a human had left the ground but it was the first time one had done so with the freedom of a bird—from a standing start on flat ground, without the aid of a balloon. The tiny engine had propelled the vehicle forward with enough force to push air under the unique wings and lift them free of terra firma. The age of controlled flight had begun.

While the Wright brothers were making improvements to their flying machine Robert Goddard was dreaming of space travel. He would spend his entire life contemplating the problems of building a rocket—one with enough power to leave the Earth.

It was easy to calculate how much energy was needed to propel any given weight away from the Earth. The mathematics were incontrovertible. The energy requirements were substantial. If you were to drop a 20 ton rock from a 100-mile-high orbit in the sky and let it fall to Earth it would release a vast amount of energy as it collided with the air, creating a fiery plume during its descent and ultimately punching a huge hole in the ground, simultaneously releasing, on impact, another stupendous amount of energy in the form of an explosion. Evidence of the kind of damage that can be caused by such an impact can be seen in photographs taken in Tunguska Siberia where 1000 square kilometers of forest were flattened in 1908. Goddard knew that this was the same amount of energy required to put that 20 ton rock back up into its orbit 100 miles above the Earth. The prospect of building a device to release so much energy, in a controlled fashion, seemed daunting to many—impossible to most.

Goddard was not to be deterred. He calculated that there might be enough energy in a controlled series of explosions to lift his ship into space. At first he tried to build a launcher which used a series of powder cartridges, like a powerful machine gun, but the rapid firing requirements continuously overwhelmed his mechanical capabilities. Then he struck upon the same idea which had occurred to Tsiolkovsky on the other side of the world. If a flow of gas could be expelled from a nozzle in a continuous and consistent manner it might be possible to use the thrust to leave the Earth on the top of a column of hot gasses. The next thing to contemplate was the fuel. It was known at a very early stage that a mixture of hydrogen and oxygen provided the most efficient solution.

Because any spacecraft would be required to also lift its own fuel the best choice was hydrogen because it is also the lightest. But hydrogen is also very bulky, so Goddard settled for mixing alcohol with oxygen. Over the next twenty years Goddard worked in almost total solitude, using money provided by the Smithsonian Institute and the Guggenheim Foundation. Some of his findings were published in Germany and were eagerly consumed by the young members of a civilian rocket club called the VFR.

The VFR had been formed less than one year after Goddard flew his first liquid fueled rocket (which took place March 16th 1926). The club included an assortment of young engineering enthusiasts whose stated goal was to send humans into space. They had all read a book by the Transylvanian, Oberth, written in 1923, called *Die Rakete* (The Rocket). Oberth had written to Goddard in the Spring of 1922 and Goddard had sent Oberth his paper. The following year Oberth published his findings which spurred Goddard to publicly dispute Oberth's precedence. Regardless of all this, the ideas for liquid fuel rockets slowly began to take shape as actual hardware. It would be five years after Goddard's first flight that the Germans would realize a successful launch to an altitude of 60 metres. One of the young enthusiasts at the VFR was the son of a German baron, his name was Wernher von Braun and he was quickly recognized as one of Oberth's most brilliant students. Von Braun had also been inspired by Lasswitz and Verne as a child.

At the end of 1930 a new facility was established for launching rockets in Germany and within two years von Braun was able to demonstrate a liquid fuel rocket for his sponsors in the German army. Less than a year later Adolf Hitler would become chancellor of Germany and von Braun's work would be placed on a more sinister path.

World War II was the catalyst that enabled the development of increasingly powerful rockets. Designed on the Baltic coast and built in ever larger quantities in underground factories, the liquid propelled rockets of Wernher von Braun would blossom into a full fledged strategic ballistic missile. It was called the A-4 but Hitler would point it at the cities of Europe and rename it the V-2.

Over 3500 of these deadly weapons were launched during the war against military and civilian targets but they were not the only rocket-propelled weapons to wreak havoc. Goddard's own "bazooka" would also claim many lives, as would the Soviet *Katusha* rocket, but the V-2 could fly at supersonic speeds and almost always arrived unannounced.

As it became inevitable that Germany would lose the war, von Braun assembled his team and they unanimously decided to surrender to the approaching American forces. In a compelling legend of subterfuge and trickery the Americans managed to not only capture the rocket team but also were able to infiltrate behind the lines of the advancing Soviet forces and spirit away entire train-loads of V-2 components and completed missiles.

In 1946 a young Ukrainian named Sergei Korolev was pardoned from some trumped-up charges and released from his prison cell. Korolev had been one of the early Russian rocket enthusiasts. He had been involved with Russia's own rocket clubs in the 1930's but he had been wrongly condemned to a prison sentence by the paranoid regime of Joseph Stalin. On his release he was commissioned as a Major in the Red Army and sent to Germany. He arrived at the remains of the German rocket site at Peenemünde and was instructed to do his very best to acquire as much useful hardware and personnel as possible. He then arranged to have these acquisitions shipped back to Russia where he would investigate their potential.

The Soviets were as astounded as the Americans had been by the enormous advance in technology represented by the V-2. Von Braun and his team had managed to build a rocket with over 20 tons of thrust; by comparison Goddard had never managed more than 7,500 pounds. It was soon apparent that the German technology had great potential as a weapon and both the Soviets and the Americans began frenzied research programs to see if the V-2 could be improved.

Von Braun and over a hundred of his German colleagues were incarcerated in the desert in New Mexico while Helmut Grottrup and many of the Germans who remained at Peenemünde were pulled out of their homes in the middle of the night and taken deep into Soviet Russia. While Korolev was reluctant to employ the Germans—since he had been working on rockets for many years before the war—the Americans had captured the team leader and were quick to recognize and employ the young genius.

A race soon began and both teams struggled to be the first to understand and employ the extraordinarily powerful new German technology.

In 1956 the United States and the Soviet Union both announced their intention to launch an earth-orbiting artificial satellite during the International Geophysical Year of 1957. At the time many politicians on both sides saw little point in spending the money on such a wasteful extravagance but meanwhile the military leaders continued to perfect larger and larger launch vehicles. Then on October 4th 1957 Korolev made his move.

Behind the Soviet curtain of secrecy he had pushed the German innovations to their limit and combined them with his own considerable expertise. Because the early Soviet designs for hydrogen bombs were much larger and cumbersome than their American equivalents he had been forced to build a much more powerful launch vehicle. The R-7 rocket included multiple combustion chambers with each firing through a set of four nozzles. It was a major accomplishment and on that cold autumn night in October the world looked up with awe at the tiny metal sphere, known as Sputnik, as it passed overhead at almost 25 times the speed of sound. It was a major political triumph for the Soviet Union and one which the Kremlin barely even recognized—at first.

When the world's plaudits began to pour into Moscow, Nikita Kruschev suddenly realized the importance of Korolev's accomplishment. He swiftly ordered that another satellite be launched in time for an impending political anniversary and Korolev obliged by sending up Sputnik 2. It was even larger than Sputnik 1 and considerably heavier.

Back in Washington nerves were frayed. Accurate intelligence of what had been going on behind the Iron Curtain was nowhere near as accurate as it is today—after all, there were no spy satellites. The implications were profound. The ability to fling a huge payload into space, such as Sputnik 2, meant that the Soviets also had the ability to send a nuclear weapon anywhere on Earth. America absolutely had to catch up.

By this time Wernher von Braun had become firmly ensconced into American society. In 1956 he had tried to convey to his superiors that he could have launched a satellite, but his pleas had fallen on deaf ears. Just as the Soviets had not wanted to let the Germans be able to claim their accomplishment it seems that there may have been similar prejudices in America. Much of the American effort had gone into a Navy program which became known as *Project Vanguard*. It was an untried and untested missile which employed many new technologies. In an attempt to show that America had not fallen too far behind Vanguard was prepared for launch. On December 6th 1957 the die was cast and the untried rocket billowed to life—but it was not to be.

Moments after leaving the launch pad the Vanguard settled back downwards and disappeared into an enormous fireball. The next day the failure was ridiculed both by the Western media and by the Soviet Premier. Von Braun once again urged the government to let him try his latest derivative of the V-2, known as the Redstone missile. It was much smaller than Korolev's R-7 but von Braun knew it was capable of placing a small satellite in orbit. With the failure of Vanguard it was decided to give von Braun a green light.

On January 31st 1958 the Redstone fulfilled von Braun's predictions and successfully placed the tiny *Explorer 1* satellite into low earth orbit, the space race had begun and America was running second in a two horse race.

What was not known in the West was just how pressed Korolev and his team were. To maintain the fast pace they had established with Sputnik 1 and 2 was a daunting prospect. Korolev was having to fulfill the roles which in America were performed by many different people—but he was relentless.

Over the next few years it seemed that America would never catch up as the Soviets continued to chalk-up a string of impressive accomplishments. The first animal in orbit, the first satellite in solar orbit, the first satellite around the moon and the first to impact on the moon. Korolev even had to devise innovative methods to build and launch communication satellites even though the Soviet launch facilities were geographically placed in such a way as to make it much harder than launching from Florida. Despite the lack of support and funding, on April 12th 1961, a handsome young 27 year old fighter pilot from Klushino outside Moscow became the first human in space. His name was Yuri Gagarin and his flight was Korolev's greatest achievement.

America's newly appointed astronaut team were tremendously disappointed, especially Alan Shepard who had believed that the newly created American manned spacecraft had been ready to take him into space a month earlier than Gagarin. America's spacecraft was christened *Project Mercury* and the small one-man ship was the result of tireless research by the West's top engineers. The team assembled at the newly formed National Aeronautics and Space Administration comprised of some of the best engineers from America, Germany, Canada and England. *Mercury* had been designed using the latest results derived from launching nuclear warheads into space. As these warheads had been lofted above the atmosphere it was discovered that if they came back sharp-end first, they melted. An enthusiastic team of engineers concluded that the best way to remove all of that energy from a returning spacecraft was to bring it back blunt-end first, using the friction of the air to gradually slow the spacecraft as it bled away the enormous velocity.

The idea worked and after much research (and some test flights with chimpanzees) only two weeks after Gagarin's triumph, Alan B. Shepard flew a short 15 minute flight into space inside his *Freedom 7* Mercury capsule. He was launched on von Braun's *Redstone* missile. On May 25th, only twenty days after Shepard's successful return from space, President John Fitzgerald Kennedy made a momentous announcement to the Congress, this is some of what he said,

"Recognizing the head start obtained by the Soviets with their large rocket engines, which gives them many months of lead-time, and recognizing the likelihood that they will exploit this lead for some time to come in still more impressive successes, we nevertheless are required to make new efforts of our own. For while we cannot guarantee that we shall one day be first, we can guarantee that any failure to make this effort will make us last. We take an additional risk by making it in full view of the world, but as shown by the feat of astronaut Shepard, this very risk enhances our stature when we are successful.

"First, I believe that this nation should commit itself to achieving the goal, before this decade is out, of landing a man on the Moon and returning him safely to the Earth. No single space project in this period will be more impressive to mankind, or more important for the long-range exploration of space; and none will be so difficult or expensive to accomplish. We propose to accelerate the development

of the appropriate lunar spacecraft. We propose to develop alternate liquid and solid fuel boosters, much larger than any now being developed, until certain which is superior. We propose additional funds for other engine development and for unmanned explorations—explorations which are particularly important for one purpose which this nation will never overlook: the survival of the man who first makes this daring flight. But in a very real sense, it will not be one man going to the Moon—if we make this judgement affirmatively, it will be an entire nation. For all of us must work to put him there."

Ironically, even though the President (and everyone else) thought the Soviets had a large lead in propulsion, one of the great advantages which the American's actually had over the Soviets was that they had started work a few years earlier on a gigantic rocket engine known as the F-1. If it could be built it would marshall a staggering 1.5 million pounds of thrust. Almost forty times as powerful as von Braun's V-2 engine.

While Korolev was mired in political infighting with his rivals in Russia, Wernher von Braun was forging ahead with his new series of launch vehicles called *Saturn*. Now that the President had made it a national priority to send Americans to the moon the money began to flow into the newly established *Apollo* program. Taking advantage of the lead in research on large engines, von Braun began to design his masterpiece.

It was to be a machine of unparalleled proportions. It would have to be able to lift many tons of hardware, not only into space, but out of Earth orbit and down to the surface of the moon. Initially discussions took place to use a launcher called *Nova*. There were many versions of *Nova,* each from different contractors, but the largest would have used eight F-1 engines and would have been powerful enough to send a manned vehicle straight to the moon and land feet first—like *Buck Rogers*.

That plan was shelved in favor of a less cumbersome and more elegant solution called *Lunar Orbit Rendezvous (LOR)*. This method would require a smaller launch vehicle but it would presume that two spacecraft could meet each other in orbit around the moon. If LOR could be made feasible, the five engined C-5 rocket would suffice to make the whole thing work. A decision was quickly taken and the course was set.

Apollo would be launched with three modules atop of a C-5 rocket, later redubbed the *Saturn V*. The three modules would be a Command Module (CM) in which the crew would spend much of their time, a Service Module (SM) which would remain attached to the CM for much of the flight and carry fuel, water, power supplies and oxygen, and a Lunar Module (LM) which would carry two of the three-man crew down to the lunar surface.

But much remained to be done to undertake such a momentous journey. Spacesuits needed to be designed that could function in the harsh vacuum of space and protect the astronaut from radiation and temperature extremes. It would be necessary to invent new ways of tracking the vehicles in flight so that they knew where they were in space. Then it would be imperative that once the two spacecraft found each other that they could maneuver together and actually connect. These were very complex issues which would not be solved quickly or cheaply and they had to be done while Apollo was still being constructed.

It was therefore decided to create a larger version of the Mercury spacecraft which could hold two astronauts and could be used to train the crews in the complexities of spacial navigation. It was called *Gemini* and between 1965 and 1966 it would fly a dozen missions, ten of them with people on board. *Project Gemini* became the *Cinderella* story of the space race—overshadowed by the glamorous Apollo—the two-seater spacecraft would perform almost flawlessly and during the course of the program it would allow America to keep up with the Soviets and ultimately surpass them.

Gemini astronauts space-walked and they rendezvoused. They used new spacesuits and new power systems and they learned how to dock. They would also establish endurance records by remaining in the cramped capsule for the full duration of a potential trip to the moon. Gemini was an unbridled success and it would be the turning point in the race.

Meanwhile, in the Soviet Union, Korolev had done his best to consistently impress the Politburo and the Americans. He had been the first to fly a woman in space, the first to fly a three man spacecraft and the first to fly two manned spacecraft at the same time, but there was only so much he could do. His design for a giant lunar rocket was sitting unrealized and embroiled in bitter infighting. Then in 1966 what was to have been a simple surgery to remove some intestinal polyps spiralled out of control and, despite the best efforts of the surgical team, Korolev died at 65 years old. No one in the west knew it yet but the space race was already over.

Not everything was well in the American camp. While Gemini was routinely over-performing in Earth orbit, the Apollo Command Module was bogged down in design problems. One of the most serious mistakes had been made years earlier when it had been decided to use a pure oxygen atmosphere inside the spacecraft. The choice had been made early on with

Mercury, and had continued through Gemini into Apollo. If you used pure oxygen inside the ship you didn't need to have such high internal pressure to give the crew the same amount of oxygen you get here at sea-level on Earth. You also only needed to maintain a single gas supply. This meant that the ship could be built lighter and thus could be lofted into space by smaller rockets. The only problem is that pure oxygen changes the way that materials behave. The slightest spark in pure oxygen can even make things that don't normally burn explode like a bomb.

A year and two weeks after the Soviets lost Korolev three American astronauts were sitting inside the first Apollo capsule atop a smaller version of von Braun's Saturn rocket. They were Virgil Grissom, Edward White and Roger Chaffee. Grissom was America's second man in space and White had been America's first space-walker, Chaffee was new to space flight.

The inside of the Apollo spacecraft was plastered in the new darling of the space-age—velcro. The astronauts had kept asking for more to be placed in the vehicle because it was so convenient for keeping things in place once you were in space. What no one realized was that a wire had lost its insulation somewhere in the base of the cabin and at precisely the right moment was waiting to send a spark into the pure oxygen atmosphere. When that particular circuit was closed the spark raced up through the cabin and in an instant it ignited the velcro and everything else within reach. In less than thirty seconds the crew were asphyxiated. The pressure caused by the fire made it impossible to open the hatch, which opened inwards. During the subsequent investigation astronaut Frank Borman said that he and his colleagues at NASA had failed due to a lack of imagination. No one foresaw any danger while the spacecraft was on the ground. The Apollo capsule had been a giant oxygen bomb waiting to go off.

If anything could have derailed Kennedy's train it was the loss of Apollo 1, but the NASA team were dauntless and determined not to lose their compatriots in vain. Apollo was redesigned and forged anew. The program pushed ahead.

Over the next 21 months a series of unmanned tests were conducted on both the Apollo spacecraft and the launch vehicles. On November 9th 1967 von Braun's giant Saturn V rocket finally flew away from the new launch complex at the Kennedy Space Center. It carried an unmanned Apollo command module which would be dubbed *Apollo 4*. Five months later it would fly again, this time with *Apollo 6*. Then in October 1968 the newly redesigned Apollo spacecraft was ready for its first manned flight. It was to be called *Apollo 7* and would fly atop the *Saturn I-B* rocket. The crew would be commanded by Mercury and Gemini veteran Wally Schirra accompanied by two rookies, Walt Cunningham and Donn Eisele. Their mission was to stay in Earth orbit and see if the Apollo was able to function for two weeks in space. The flight went off without a hitch and prepared the way for the most hectic nine months in the history of spaceflight.

Just before the flight of Apollo 7 it had been learned from intelligence sources that the Soviets were building a giant moon rocket. After Korolev's death the enormous N-1 booster had somehow survived the machinations of opposing camps and was on its way to the launch pad. The N-1 stood almost as tall as the Saturn V but it did not have the advantage of the stupendously powerful F-1 engines that drove the Saturn V. Another advantage that the Americans had was the extremely successful program to perfect large hydrogen fueled rockets for the upper stages of the Saturn. The N-1 was the Soviets last gambit and the implications of losing the race, at this late stage, was unthinkable to the American team. The manager of the Apollo program, Dr George Mueller, had pushed to test the Saturn V all in one go, or "all-up" as it was called. This had worked brilliantly but now it was seemingly necessary to push even harder. What Mueller and his team couldn't have known was that the N-1 was fundamentally flawed and was destined to disappear in a fireball each and every time it was launched. Without an equivalent to the F-1 engine the N-1 was simply too hard to handle, requiring 30 engines on its first stage. The logistics of sequencing such a large number of engines was plainly too much for the comparatively old Soviet computers.

The American method of reaching the moon called Lunar Orbit Rendezvous required that the spacecraft use a detachable lander module which would fly down to the lunar surface while the main spacecraft remained in orbit. Unfortunately the lander was not ready and the next Saturn V was sitting idle awaiting its payload.

The contractor building the LM had encountered more and more problems with the delicate spider-like lander and couldn't deliver the first one in time. So a risky decision was taken. Send the next Apollo around the moon without a lander. It would mean that all of the systems could be checked out, including the tracking and communications but no one would land. It would also have the added advantage of upstaging the Soviets. So on December 21st 1968 a crew comprising of Frank Borman, James Lovell and William Anders found themselves sitting on top of a Saturn V awaiting their moment in history. At 7.50 in the morning the tension was palpable as the world watched five F-1 engines roar to life and slowly move the massive rocket into the air. Everything performed flawlessly and within a few hours the trio were leaving the Earth behind. It was the first time anyone had ever left Earth orbit and headed into open space. Less than a day into the voyage of Apollo 8 the crew turned on a television camera and showed the Earth receding away from them.

On Christmas Eve Borman, Lovell and Anders radioed the people back home and read extracts from the Bible. They flew in orbit a mere sixty miles above the lunar surface and saw things which no human eyes had ever beheld. The far side of the moon, obscured from human sight since the dawn of time revealed its secrets to Apollo 8, the Earth rose above the lunar horizon and the crew scrambled for the cameras. Entire continents were visible at a glance.

Back on Earth the teams at NASA and Grumman scrambled to meet deadlines. The clock was ticking and the first Lunar Module still wasn't ready. Apollo 8 would return to Earth exactly as planned and would land precisely where it was supposed to, in the Pacific Ocean, thus fulfilling Jules Verne's predictions in the most uncanny way possible. Verne had flown his ship from Florida, around the moon and back to splashdown in the Pacific, all in his imagination, back in 1865.

On the movie screens around the world people were engrossed with Stanley Kubrick's motion picture *2001 A Space Odyssey*. It depicted a mysterious and sterile future for humanity in space and was in some respects a harsh contrast to the reality of Apollo, but it also depicted a future of optimism where humankind would one day build cities on the moon and reach even further afield with titanic nuclear-propelled vessels, like those mentioned in Kennedy's speech of 1961. The reality of 1968 was somewhat more pedestrian for those working the late nights, but no less exciting for a generation of avid young space enthusiasts who gathered around televisions every night to hear the latest developments in this greatest of quests.

In March of 1969, with less than nine months left in Kennedy's decade the world's first true spacecraft took flight in low Earth orbit. It was nick-named *Spider* and it earned the name. Constructed from the most delicate design and conservative use of materials the Lunar Module of Apollo 9 danced its way around the Earth every ninety minutes. Launched again on a Saturn V, this time with James McDivitt, David Scott and Russell Schweickart, Apollo 9 was the final test in Earth orbit before sending the entire Apollo system to the moon.

McDivitt and Schweickart detached the lunar module from the command module and flew it away. This was only the second time that NASA had flown two manned vehicles in space simultaneously and this time it was with a vehicle that had never flown before. It was a risky business and involved a string of "firsts" to complete the mission. Schweickart was scheduled to don his lunar spacesuit and step outside. This was the first time anyone had ever done this in an Apollo spacesuit and it was complicated by the fact that Schweickart was ill. Then the two ships had to separate and fly away from each other. This had never been done before with an untested vehicle and finally the LM was to be separated from its descent stage before seeking out the command module and redocking. The flight plan for Apollo 9 was packed with hazardous and demanding tasks but it was all essential to give the timeline any chance of succeeding before the end of the year. Once again the crew and vehicles performed flawlessly and after ditching *Spider* into the Earth's atmosphere they returned safely, setting the stage for the final all-up test of Apollo 10.

Exactly eight years to the day after President Kennedy made his request to Congress an all-veteran crew comprising Thomas Stafford, John Young and Eugene Cernan were on their way home after completing the final test flight of Apollo. They had piloted their craft across the 240,000 mile gulf and taken their lunar module down to within 47,000 feet of the lunar surface. They had flown closer to the moon than many of them had flown high performance fighter jets back on Earth. Even then, things were not quite as simple as they may have seemed. At the last minute, before detaching the descent stage, one of the controls was inadvertently left in the wrong setting and when the command was given to detach, the ascent stage, with the crew aboard, flew into an extremely hazardous spin. It was referred to in the post flight report as "large attitude excursions" but for a moment it seriously jeopardized the crew's well-being. Stafford and Cernan displaying the calm comportment drilled into them over their years of training, brought the vehicle back under control and succeeded in rendezvousing and docking. It was the first time anyone had undertaken such a complex procedure so far away from the Earth. On May 26th the crew of Apollo 10 returned home having been tantalizingly close to the final prize.

All the data were analyzed carefully. Details were discussed regarding the right time of day to attempt a landing and when the shadows would be most advantageous to enable the crew to see any hazards on the surface. Gravitational anomalies had been discovered by Apollo 10 that could potentially throw off the trajectory of an orbiting vehicle. The complexities continued to build. A final landing site had been chosen in the Sea of Tranquility and the crew rotation had cycled around to three men who were the ones fortunate enough to go for broke. The stage was now set for the final push. Apollo 11 was ready to go.

THE CREW OF APOLLO 11

NAME: Neil A. Armstrong (Mr.)
NASA Astronaut,   Commander,
Apollo 11

BIRTHPLACE AND DATE: Born in
Wapakoneta, Ohio, on August 5. 1930;
he is the son of Mr. and Mrs. Stephen
Armstrong of Wapakoneta.

EDUCATION: Attended secondary
school in Wapakoneta, Ohio; received a Bachelor of Science
degree in Aeronautical Engineering from Purdue University in
1955. Graduate School - University of Southern California.

OTHER ACTIVITIES: His hobbies include soaring (for which
he is a Federation Aeronautique Internationale gold badge
holder).

ORGANIZATIONS: Associate Fellow of the Society of
Experimental Test Pilots; associate fellow of the American
Institute of Aeronautics and Astronautics; and member of the
Soaring Society of America.

SPECIAL HONORS: Recipient of the 1962 Institute of
Aerospace Sciences Octave Chanute Award; the 1966 AIAA
Astronautics Award; the NASA Exceptional Service Medal; and
the 1962 John J. Montgomery Award.

NAME: Michael Collins (Lieutenant Colonel, USAF)
NASA Astronaut, Command Module Pilot, Apollo 11

BIRTHPLACE AND DATE: Born in Rome, Italy, on October 31,

1930. His mother, Mrs. James L. Collins, resides in Washington, D.C.

EDUCATION: Graduated from Saint Albans School in Washington, D.C.; received a Bachelor of Science degree from the United States Military Academy at West Point, New York, in 1952.

OTHER ACTIVITIES: His hobbies include fishing and handball.

ORGANIZATIONS: Member of the Society of Experimental Test Pilots.

SPECIAL HONORS: Awarded the NASA Exceptional Service Medal, the Air Force Command Pilot Astronaut Wings, and the Air Force Distinguished Flying Cross.

NAME: Edwin E. "Buzz" Aldrin, Jr. (Colonel., USAF) NASA Astronaut, Lunar Module Pilot, Apollo 11

BIRTHPLACE AND DATE: Born in Montclair, New Jersey, on January 20, 1930, and is the son of the late Marion Moon Aldrin and Colonel (USAF Retired) Edwin E. Aldrin, who resides in Brielle, New Jersey.

EDUCATION: Graduated from Montclair High School, Montclair, New Jersey; received a Bachelor of Science degree from the United States Military Academy at West Point, New York, in 1951 and a Doctor of Science degree in Astronautics from the Massachusetts Institute of Technology in 1963; recipient of an Honorary Doctorate of Science degree from Gustavus Adolphus College in 1967, Honorary degree from Clark University, Worchester, Mass.

OTHER ACTIVITIES: He is a Scout Merit Badge Counsellor and an Elder and Trustee of the Webster Presbyterian Church. His hobbies include running, scuba diving, and high bar exercises.

ORGANIZATIONS: Associate Fellow of the American Institute of Aeronautics and Astronautics; member of the Society of Experimental Test Pilots, Sigma Gamma Tau (aeronautical engineering society), Tau Beta Pi (national engineering society), and Sigma Xi (national science research society) and a 32nd Degree Mason advanced through the Commandery and Shrine.

SPECIAL HONORS: Awarded the Distinguished Flying Cross with one Oak Leaf Cluster, the Air Medal with two Oak Leaf Clusters, the Air Force Commendation Medal, the NASA Exceptional Service Medal and Air Force Command Pilot Astronaut Wings, the NASA Group Achievement Award for Rendezvous Operations Planning Team, an Honorary Life Membership in the International Association of Machinists and Aerospace Workers, and an Honorary Membership in the Aerospace Medical Association.

THE FLIGHT OF APOLLO 11

PRELAUNCH

The Apollo 11 countdown was accomplished with no unscheduled holds.

LAUNCH AND EARTH PARKING ORBIT

The Apollo 11 space vehicle was successfully launched from Kennedy Space Center, Florida, at 9:32 a.m. eastern daylight time (EDT) on 16 July 1969. All launch vehicle stages performed satisfactorily, inserting the S-IVB/spacecraft combination into an earth parking orbit of 103 nautical miles (NM) circular—precisely as planned. All systems operated satisfactorily.

Pre-TLI (translunar injection) checkout was conducted as planned and the second S-IVB burn was initiated on schedule. All systems operated satisfactorily and all end conditions were nominal for the translunar coast on a free return trajectory.

TRANSLUNAR COAST

The Command/Service Module (CSM) was separated from the remainder of the orbital vehicle at about 3:17 GET, (hr:min ground elapsed time). The crew reported at 3:29 GET that CSM transposition and docking with the Lunar Module (LM)/Instrument Unit (IU)/S-IVB were complete. Ejection of the CSM/LM was successfully accomplished at about 4:17 GET and a SM Service Propulsion System (SPS) evasive maneuver was performed as planned at 4:40 GET. All launch vehicle safing activities were performed as scheduled.

The S-IVB/IU slingshot maneuver was successful in avoiding spacecraft recontact, lunar impact and earth capture. The closest approach to the moon of 2340 nautical miles occurred at 78:50:34 (4:22:34 p.m. EDT, 19 July).

The accuracy with which the Trans Lunar Injection (TLI) maneuver was performed was such that midcourse correction number one (MCC-1) was not required.

An unscheduled 16-minute television transmission was recorded at the Goldstone station beginning at 10:32 GET. The tape was played back at Goldstone and transmitted to Houston beginning at 11:26 GET. An unscheduled 50-minute television transmission was accomplished at 30:28 GET, and a 36-minute scheduled TV transmission began at 33:59 GET.

MCC-2 was performed at 26:45 GET as planned and all SPS burn parameters were nominal. The accuracy of the MCC-2 calculation and performance was such that MCC-3 and MCC-4 were not necessary.

The crew initiated a 96-minute color television transmission at 55:08 GET. The picture resolution and general quality were exceptional. The coverage was outstanding, including the interior of both the CM and LM, and views of the exterior of the CM and the earth. Excellent views of the crew accomplishing probe and drogue removal, spacecraft tunnel hatch opening, LM housekeeping and equipment were obtained. The picture was transmitted live throughout North and South America, Japan, and Western Europe.

The spacecraft passed into the moon's sphere of influence at 61:39:55 GET. At that time, the distance from the spacecraft to earth was 186,436 nautical miles (NM) and its distance from the moon was 33,823 NM. The velocity was 2,990 feet per second (fps) relative to earth and 3,775 fps relative to the moon.

LUNAR ORBIT

Lunar orbit insertion (LOI) was performed in two separate maneuvers using the SPS. The first maneuver, LOI-I, was initiated at 75:50 GET. The retrograde maneuver placed the spacecraft in a 168.8 x 61.3 NM elliptical orbit.

During the second lunar orbit, a scheduled live color television transmission was accomplished. Spectacular views of the lunar surface included the approach path to Lunar Landing Site 2.

After two revolutions and a navigation update, a second SPS retrograde burn (LOI-2) was made. The resulting orbit had an apolune (high point) of 65.7 NM and a perilune (low point) of 53.8 NM. LOI-I and 2 burn parameters were nominal.

After LOI-2, the crew transferred to the LM and for about two hours performed various housekeeping functions, a voice and telemetry test, and an Oxygen Purge System check. LM functions and consumables quantities checked out very well. Additionally, both LM Hasselblad and Maurer cameras were checked and verified as being operational.

The Commander (CDR) and LM Pilot (LMP) re-entered the LM at approximately 95:20 GET to perform a thorough check of all LM systems in preparation for descent. Undocking of the LM from the CSM occurred at approximately 100:14 GET. Station-keeping was then initiated. At 100:40 GET, the SM Reaction Control System (RCS) was used to perform a small separation maneuver directed radially downward toward the center of the moon as planned.

DESCENT

The descent orbit insertion (DOI) maneuver was performed by a LM Descent Propulsion System (DPS) retrograde burn one-half revolution after LM/CSM separation, placing the LM in an elliptical orbit whose perilune was 8.5 NM.

The LM powered descent maneuver was initiated at perilune of the descent orbit. The time of powered descent initiate (PDI) was as planned. However, the position at which PDI occurred was about 4 NM downrange from that which was expected. This resulted in the landing point being shifted downrange about 4 NM.

During the final approach phase, the crew noted that the landing point toward which the spacecraft was headed was in the center of a large crater which appeared extremely rugged with boulders of 5 to 10 feet in diameter and larger. Consequently, the crew elected to fly to a landing point beyond this crater. This required manual attitude control and fine adjustments of the rate of descent plus high horizontal velocity to translate beyond the rough terrain area. As indicated above, the final landing point was estimated to be about four miles downrange from the center of the planned landing ellipse. Lunar landing occurred at 102:45:43 GET (4:17:43 P.M. EDT). Current estimate of landing point coordinates, based on analysis of all available data is 23.47° E and .67° N. This is approximately 20,800 feet west and 4,000 to 5,000 feet south of the center of the planned landing ellipse. Site altitude is estimated at approximately 8600 ft below the moon's mean radius.

LUNAR SURFACE ACTIVITIES

LM attitude on the surface was tilted 4.5° from the vertical, and yawed left about 13°. The crew indicated the landing site area contained numerous boulders of varying shapes and sizes up to 2 feet. The surface color varied from very light to dark gray. From his window view, the CDR reported seeing some boulders that were apparently fractured by engine exhaust. He reported that the surface of these boulders appeared to be coated light gray whereas the fractures were much darker. A hill was in sight at about ½ to 1 mile in front of the LM.

The crew indicated that they could immediately adapt to the 1/6 (earth) gravity in the LM and moved very easily in this environment. About two hours after landing, the crew requested that the extravehicular activity (EVA) be accomplished prior to the sleep period, or about 4½ hours earlier than originally scheduled. The rest period originally planned to occur prior to EVA was slipped until post-EVA and added to the second sleep period.

EXTRAVEHICULAR ACTIVITY

After the postlanding checks, the LM hatch was opened at 109:07:35 GET. As the CDR descended the LM ladder, he deployed the Modularized Equipment Stowage Assembly (MESA). The television camera mounted on the MESA access panel recorded his descent to the lunar surface. The CDR's first stop on the moon occurred at 109:24:15 GET (10:56:15 p.m. EDT). He made a brief check of the LM exterior, indicating that penetration of the footpads was only about 3 to 4 inches and collapse of the strut was minimal. He reported sinking approximately 1/8 inch into the fine, powdery surface material, which adhered readily to the lunar boots in a thin layer. There was no crater from the effects of the descent engine, and about one foot of clearance was observed between the engine bell and the lunar surface. He also reported that it was quite dark in the shadows of the LM which made it difficult for him to see his footing. During the EVA, a small microdot disk containing messages from numerous world leaders was left on the moon.

The CDR then collected a contingency sample of lunar soil from the vicinity of the LM ladder. He reported that although loose material created a soft surface, as he dug down six or eight inches, he encountered very hard cohesive material.

The CDR photographed the LMP's egress and descent to the lunar surface. The CDR and LMP then unveiled the plaque mounted on the strut behind the ladder and read its inscription to their worldwide television audience. Next, the CDR removed the TV camera from the Descent Stage MESA, obtained a panorama, and placed the camera on its tripod in position to view the subsequent surface EVA operations.

The LMP deployed the Solar Wind Composition experiment on the lunar surface in direct sunlight to the north of the LM as planned.

Subsequently, the crew erected a 3 x 5-foot American flag on an 8-foot aluminum staff. During the ensuing environmental evaluation, the LMP indicated that he had to be careful of his center of mass in maintaining balance. He noted that the LM shadow had no significant effect on his EMU temperature. The LMP also noted that his agility was better than expected. He was able to move about with great ease. Both crewmen indicated that their mobility throughout the EVA significantly exceeded all expectations. Also, indications were that metabolic rates were much lower than premission estimates.

A conversation between President Nixon and Armstrong and Aldrin was held. The conversation originated from the White House and contained congratulations and good wishes.

The CDR collected a Bulk Sample consisting of assorted

surface material and selected rock chunks, and placed them in a Sample Return Container (SRC). Following the Bulk Sample collection, the crew inspected the LM and reported no discrepancies. The quads, struts, skirts, and antennas were satisfactory.

The Passive Seismic Experiment Package (PSEP) and Laser Ranging Retro Reflector were deployed south of the LM. Excellent PSEP data was obtained including detection of the crewmen walking on the surface and later, in the LM. The crew then collected more lunar samples until EVA termination including two core samples and about 20 pounds of discretely selected material. The LMP had to exert a considerable force to drive the core tubes an estimated six to eight inches deep.

Throughout the EVA, TV was useful in providing continuous observation for time correlation of crew activity with telemetered data and voice comments and in providing live documentation of this historically significant achievement. Lunar surface photography consisted of both still and sequence coverage using the Hasselblad camera, the Maurer data acquisition camera and the Apollo Lunar Surface Close-Up camera.

EVA termination film and sample transfer, LM ingress, and equipment jettison were accomplished according to plan. A rest period followed the post-EVA activities prior to preparation for liftoff.

ASCENT, RENDEZVOUS AND TRANSEARTH INJECTION

LM liftoff from the lunar surface occurred at 124:22 GET (1:34 p.m. EDT, 21 July) concluding a total lunar stay time of 21 hours 36 minutes. All lunar ascent and rendezvous maneuvers were nominal and terminated with CSM/LM docking at 128:03

GET. After transfer of the crew, samples and film to the CSM, the LM Ascent Stage was jettisoned at 130:10 GET. The LM ascent stage will remain in lunar orbit for an indefinite period of time. Subsequently, a small SM RCS separation maneuver placed the CSM in a 62.6 by 54.7 nautical mile orbit.

At 135:24 GET, the SPS injected the CSM into a transearth trajectory after a total time in lunar orbit of 59 hours 28 minutes (30 revolutions). The TEI resulted in a transearth return time of about 60 hours.

TRANSEARTH COAST AND ENTRY

MCC-5 was initiated at 150:30 GET. The 10.8 second SM RCS burn produced a velocity change of 4.7 feet per second. An 18-minute television transmission was initiated at 155:36 GET and produced good quality pictures. The transmission featured crew demonstrations of the effect of weightlessness on food and water and brief scenes of the moon and earth. The accuracy of MCC-5 was such that MCC-6 and MCC-7 were not required.

The final color television broadcast was made at 177:32 GET. The 12½ minute transmission featured a sincere message of appreciation by each crew member to all people who helped make the Apollo 11 mission possible.

The crew awoke at 189:15 GET and initiated reentry preparations. CM/SM separation occurred at 194:49:19 GET and entry interface was reached at 195:03 GET.

Because of deteriorating weather in the nominal landing area, the aim point had been moved downrange 215 NM. Weather in the new landing area was excellent: visibility was 12 miles, wave height 3 feet, and wind 16 knots.

Visual contact of the spacecraft was reported at 195:06 GET. Drogue and main parachutes deployed normally. Landing occurred about 14 minutes after entry interface at 195:18:35 GET (12:50:35 EDT). The landing point was in the mid-Pacific, approximately 169:09°W longitude by 13:18° N latitude, about 13 NM from the prime recovery ship, USS HORNET. The CM landed in the Stable 2 position. Flotation bags were deployed to right the S/C into Stable 1 position at 195:25:10. The crew reported that they were in good condition.

EARLY APOLLO SCIENTIFIC EXPERIMENTS PACKAGE (EASEP)

The Apollo 11 scientific experiments for deployment on the lunar surface near the touchdown point of the lunar module were stowed in the LM's scientific equipment bay at the left rear quadrant of the descent stage looking forward.

The Early Apollo Scientific Experiments Package (EASEP) was carried only on Apollo 11; subsequent Apollo lunar landing missions carried the more comprehensive Apollo Lunar Surface Experiment Package ALSEP.

EASEP consisted of two basic experiments: the passive seismic experiments package (PSEP) and the laser ranging retro-reflector (LRRR). Both experiments were independent, self-contained packages that

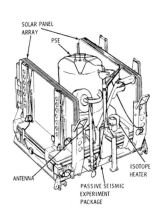

PSEP STOWED CONFIGURATION

SOLAR PANEL ARRAY
PSE
ISOTOPE HEATER
ANTENNA
PASSIVE SEISMIC EXPERIMENT PACKAGE

weighed a total of about 170 pounds and occupied 12 cubic feet of space.

PSEP used three long-period seismometers and one short-period vertical seismometer for measuring meteoroid impacts and moonquakes. Such data was useful in determining the interior structure of the Moon; for example, does the Moon have a core and mantle like Earth?

SOLAR WIND EXPERIMENT

The seismic experiment package had four basic subsystems: structure/ thermal subsystem for shock, vibration and thermal protection; electrical power subsystem which generated 34 to 46 watts by solar panel array; data subsystem received and decoded MSFN uplink commands and downlinks experiment data, handled power switching tasks; passive seismic experiment subsystem measured lunar seismic activity with long-period and short-period seismometers which detected inertial mass displacement.

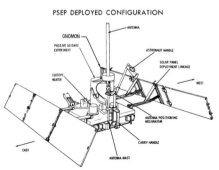

PSEP DEPLOYED CONFIGURATION

GNOMON
ANTENNA
PASSIVE SEISMIC EXPERIMENT
ASTRONAUT HANDLE
SOLAR PANEL DEPLOYMENT LINKAGE
ISOTOPE HEATER
WEST
ANTENNA POSITIONING MECHANISM
CARRY HANDLE
EAST
ANTENNA MAST

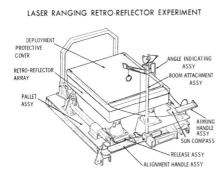

LASER RANGING RETRO-REFLECTOR EXPERIMENT

DEPLOYMENT
PROTECTIVE
COVER
RETRO-REFLECTOR
ARRAY
PALLET
ASSY
ANGLE INDICATING
ASSY
BOOM ATTACHMENT
ASSY
AIMING
HANDLE
ASSY
SUN COMPASS
RELEASE ASSY
ALIGNMENT HANDLE ASSY

The laser ranging retro-reflector experiment is a retroreflector array with a folding support structure for aiming and aligning the array toward Earth. The array is built of cubes of fused silica. Laser ranging beams from Earth have been reflected back to their point of origin for precise measurement of Earth-Moon distances, motion of the Moon's center of mass, lunar radius and Earth geophysical information.

Earth stations which beamed lasers to the LRRR included the McDonald Observatory at Ft. Davis, Tex.; Lick Observatory, Mt. Hamilton, Calif.; and the Catalina Station of the University of Arizona. Scientists in other countries also bounced laser beams off the LRRR.

Life Support Equipment - Space Suits

Apollo 11 crewmen wore two versions of the Apollo space suit: an intravehicular pressure garment assembly worn by the command module pilot and the extravehicular pressure garment assembly worn by the commander and the lunar module pilot. Both versions are basically identical except that the extravehicular version had an integral thermal/ meteoroid garment over the basic suit.

From the skin out, the basic pressure garment consists of a

nomex comfort layer, a neoprene-coated nylon pressure bladder and a nylon restraint layer. The outer layers of the intravehicular suit are, from the inside out, nomex and two layers of Teflon-coated Beta cloth. The extravehicular integral thermal/meteoroid cover consists of a liner of two layers of neoprene-coated nylon, seven layers of Beta/kapton spacer laminate, and an outer layer of Teflon-coated Beta fabric.

The extravehicular suit together with a liquid cooling garment, portable life support system (PLSS), oxygen purge system, lunar extravehicular visor assembly and other components make up the extravehicular mobility unit (EMU). The EMU provides an extravehicular crewman with life support for a four hour mission outside the lunar module without replenishing expendables. EMU total weight is 183 pounds. The intravehicular suit weighed 35.6 pounds.

Liquid cooling garment — A knitted nylon-spandex garment with a network of plastic tubing through which cooling water from the PLSS is circulated. It was worn next to the skin and replaced the constant wear-garment during EVA only.

Portable Life Support System — A backpack supplying oxygen at 3.9 psi and cooling water to the liquid cooling garment. Return oxygen is cleansed of solid and gas contaminants by a lithium hydroxide canister. The PLSS includes communications and telemetry equipment, displays and controls, and a main power supply. The PLSS is covered by a thermal insulation jacket. (Two stowed in LM).

Oxygen Purge System — Mounted atop the PLSS, the oxygen purge system provided a contingency 30-minute supply of gaseous oxygen in two two-pound bottles pressurized to 5,880 psia. The system may also be worn separately on the

EXTRAVEHICULAR MOBILITY UNIT

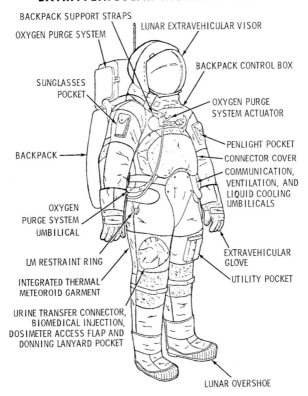

BACKPACK SUPPORT STRAPS

OXYGEN PURGE SYSTEM

LUNAR EXTRAVEHICULAR VISOR

BACKPACK CONTROL BOX

SUNGLASSES POCKET

OXYGEN PURGE SYSTEM ACTUATOR

PENLIGHT POCKET

CONNECTOR COVER

BACKPACK

COMMUNICATION, VENTILATION, AND LIQUID COOLING UMBILICALS

OXYGEN PURGE SYSTEM UMBILICAL

LM RESTRAINT RING

EXTRAVEHICULAR GLOVE

INTEGRATED THERMAL METEOROID GARMENT

UTILITY POCKET

URINE TRANSFER CONNECTOR, BIOMEDICAL INJECTION, DOSIMETER ACCESS FLAP AND DONNING LANYARD POCKET

LUNAR OVERSHOE

front of the pressure garment assembly torso. It served as a mount for the VHF antenna for the PLSS.

Lunar extravehicular visor assembly — A polycarbonate shell and two visors with thermal control and optical coatings on them. The EVA visor was attached over the pressure helmet to provide impact, micrometeoroid, thermal and ultraviolet infrared light protection to the EVA crewman.

<u>Extravehicular gloves</u> — Built of an outer shell of Chromel-R fabric and thermal insulation to provide protection when handling extremely hot and cold objects. The finger tips were made of silicone rubber to provide the crewman more sensitivity.

A one-piece constant-wear garment, similar to "long Johns", was worn as an undergarment for the spacesuit in intravehicular operations and for the inflight coveralls. The garment was porous-knit cotton with a waist-to-neck zipper for donning. Biomedical harness attach, points are provided.

During periods out of the space suits, crewmen will wore two-piece Teflon fabric inflight coveralls for warmth and for pocket stowage of personal Items.

Communications carriers ("Snoopy hats") with redundant microphones and earphones were worn with the pressure helmet; a lightweight headset was worn with the inflight coveralls.

After splashdown, the Apollo 11 crew donned biological isolation garments passed to them through the spacecraft hatch by a recovery swimmer. The crew was carried by helicopter to the Hornet where they entered a Mobile Quarantine Facility (MQF) about 90 minutes after landing. The MQF (opposite page top), with crew aboard, was offloaded at Ford Island, Hawaii

BIOLOGICAL ISOLATION GARMENT

and loaded on a C-141 aircraft for the flight to Ellington AFB, Texas, and thence trucked to the Lunar Receiving Laboratory (LRL).

THE SATURN V LAUNCH VEHICLE

OVERALL VEHICLE	DIAMETER	HEIGHT	WEIGHT
	33 ft.	364 ft.*	6,100,000 lb. (total liftoff)
FIRST STAGE	33 ft.	138 ft.	300,000 lb. (dry)
SECOND STAGE	33 ft.	81 ft. 7 in.	95,000 lb. (dry)**
THIRD STAGE	21 ft. 8 in.	58 ft. 7 in.	34,000 lb. (dry)**
INSTRUMENT UNIT	21 ft. 8 in.	3 ft.	4,500 Lb.
APOLLO SPACECRAFT		80 ft.	95,000 Lb.

*SINCE INDIVIDUAL STAGE DIMENSIONS OVERLAP IN SOME CASES, OVERALL VEHICLE LENGTH IS NOT THE SUM OF INDIVIDUAL STAGE LENGTHS
**INCLUDES AFT INTERSTAGE WEIGHT

PROPULSION SYSTEMS

FIRST STAGE - Five bipropellant F-I engines developing 7,500,000 lb. thrust RP-1 Fuel - 203,000 gal. (1,359,000 lb.), LOX-331,000 gal. (3,133,000 lb.)

SECOND STAGE - Five bipropellant J-2 engines developing more than 1,000,000 lb. thrust LH_2 - 260,000 gal. (153,000 lb.), LOX-83,000 gal. (789,000 lb.)

THIRD STAGE - One bipropellant J-2 engine developing up to 225,000 lb. thrust LH_2 - 63,000 gal. (37,000 lb.), LOX-20,000 gal. (191,000 lb.)

CAPABILITY

FIRST STAGE - Operates about 2.5 minutes to reach an altitude of about 200,000 feet (38 miles) at burnout

SECOND STAGE - Operates about 6 minutes from an altitude of about 200,000 feet to an altitude of 606,000 feet (114.5 miles)

THIRD STAGE - Operates about 2.75 minutes to an altitude of about 608,000 feet (115 miles) before second firing and 5.2 minutes to translunar injection

PAYLOAD - 250,000 Lb. into a 115 statute-mile orbit

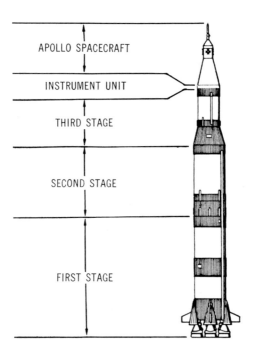

APOLLO SPACECRAFT

INSTRUMENT UNIT

THIRD STAGE

SECOND STAGE

FIRST STAGE

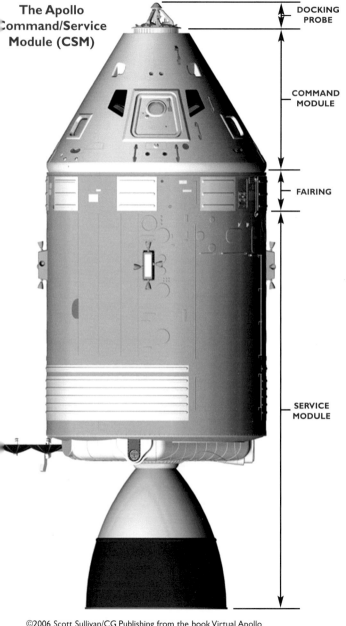

The Apollo Command/Service Module (CSM)

DOCKING PROBE

COMMAND MODULE

FAIRING

SERVICE MODULE

©2006 Scott Sullivan/CG Publishing from the book Virtual Apollo

The Apollo Command/Service Module (CSM)

TOP

BOTTOM

©2006 Scott Sullivan/CG Publishing from the book Virtual Apollo

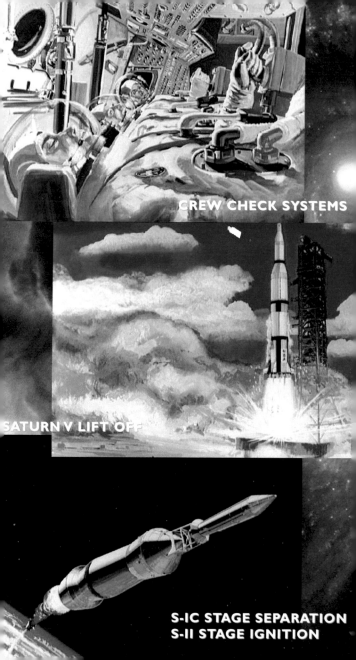

CREW CHECK SYSTEMS

SATURN V LIFT OFF

S-IC STAGE SEPARATION
S-II STAGE IGNITION

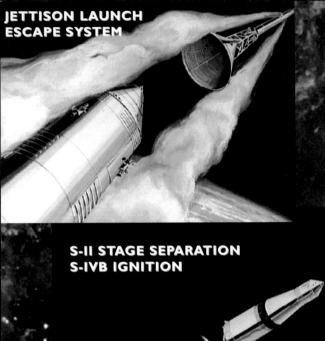

JETTISON LAUNCH ESCAPE SYSTEM

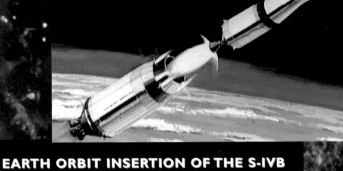

S-II STAGE SEPARATION S-IVB IGNITION

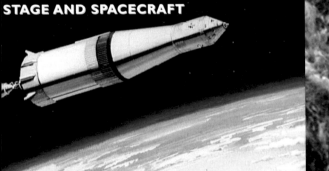

EARTH ORBIT INSERTION OF THE S-IVB STAGE AND SPACECRAFT